"The Last Forest: A Climate Change Survival Story"

By

Muhammad Hasnain Haider

Table of Contents

Chapter 1: The Journey Begins

- Introducing the main character and their world
- The inciting incident that sets them on their journey to understand climate change
- Meeting key supporting characters who will accompany them on their journey

Chapter 2: The Causes of Climate Change

- Researching the science of climate change and its causes
- Learning about the industries and systems that contribute to climate change
- Exploring the political and economic factors that make solving the problem so challenging

Chapter 3: The Consequences of Climate Change

- Witnessing the effects of climate change firsthand
- Meeting people and animals affected by climate change and learning their stories
- Wrestling with the moral and ethical implications of climate change

Chapter 4: "The Legacy Continues"

- Discovering the long history of climate activism and awareness

- Meeting older generations who have been fighting climate change for decades
- Learning from past mistakes and successes in the movement

Chapter 5: "The Future is Now"

- Developing a plan of action to address climate change
- Exploring the different ways individuals and communities can make a difference
- Struggling with the personal sacrifices and difficult choices that come with fighting climate change

Chapter 6: "The Power of Persistence: Maya and her Team's Journey to Inspire Climate Action"

- Building a team of activists and allies to join in the fight
- Overcoming obstacles and setbacks on the path to making a difference
- Celebrating small victories and staying motivated for the long fight ahead

Conclusion

- Reflecting on the journey and the lessons learned
- Recognizing that the fight against climate change is ongoing and never truly finished
- Encouraging readers to take action in their own lives and continue the fight for a sustainable future.

Short Summary

"The Last Forest: A Climate Change Survival Story"

is a science fiction novel that tells the story of a family struggling to survive in a world devastated by environmental degradation and climate change? The book depicts a future in which humanity is on the brink of extinction due to extreme weather events and rising sea levels. Through the characters, the author portrays the societal impacts of climate change and warns of the dangers of ignoring the natural world. The novel offers a thought-provoking exploration of the consequences of unchecked environmental destruction and serves as a reminder of the urgent need to address the issue of climate change.

Chapter1

The Journey Begins

The Earth was in trouble, and everyone knew it. But it wasn't until the day that Sarah stumbled upon an old journal in the attic of her grandparents' home that she truly understood the extent of the danger. The journal was filled with warnings about a phenomenon known as Climate Change, and Sarah was shocked by what she read.

She remembered her grandfather, a wise old man who had always told her to respect and care for the planet. As she flipped through the pages, she could almost hear his voice in her head. "Sarah, the Earth is a delicate balance of heat and energy. It's been that way for millions of years, but now it's changing, and it's changing fast," he had warned.

Sarah was filled with a sense of urgency. She had to do something to help. She gathered her friends, Jack and Olivia, and together they set out on a journey to uncover the truth about Climate Change.

"What exactly is Climate Change?" asked Jack as they set out.

"Well, it's the gradual increase in the Earth's temperature, caused by the trapping of heat in the atmosphere," explained Sarah. "It's caused by things like burning fossil fuels and deforestation, and it's changing the Earth's climate in ways we never imagined."

Olivia was skeptical. "I've heard about it, but I thought it was just a hoax," she said.

"No, it's not a hoax," replied Sarah. "The evidence is clear. The Earth is getting hotter, and unless we take action, the consequences will be catastrophic."

The trio traveled far and wide, speaking with scientists, activists, and everyday people. They learned about the science behind Climate Change, the impact it was having on the planet, and the ways in which people were working to mitigate its effects.

As their journey came to a close, Sarah, Jack, and Olivia had a newfound appreciation for the planet they called home. They realized that the fate of the Earth was in their hands, and they vowed to do everything they could to protect it.

"We need to spread the word and get others involved," said Sarah. "Together, we can make a difference."

And with that, the three friends set out to make a difference, determined to be a part of the solution to the greatest crisis the planet had ever faced.

As they traveled, they met many people who had been affected by Climate Change in one way or another. Farmers whose crops had been devastated by drought, coastal residents who had lost their homes to rising sea levels, and animal species that were struggling to survive as their habitats were destroyed.

"This is a crisis that affects us all, no matter where we live," said Sarah. "And it's only going to get worse unless we take action."

They also met individuals who were working tirelessly to combat Climate Change. Scientists who were researching new technologies to reduce emissions, activists who were raising awareness about the issue, and policymakers who were fighting for change at the local, national, and international level.

"It's inspiring to see so many people working together to make a difference," said Olivia. "But it's also daunting. This is a problem that's so big, it's hard to know where to start."

"We start by educating ourselves and others about the issue," said Jack. "And we support the people who are doing the hard work on the front lines. Every small action we take can add up to make a big impact."

Sarah, Jack, and Olivia returned home with a renewed sense of purpose. They were determined to use the knowledge they had gained on their journey to make a positive impact in their own communities. They held public forums, organized clean-up events, and educated their friends and family about the importance of taking action on Climate Change.

"We may not be able to solve this crisis overnight, but we can make a difference," said Sarah. "Every small action we take brings us one step closer to a future where the Earth is healthy and sustainable for generations to come."

As the trio continued their efforts, they were joined by others who were inspired by their commitment. Together, they formed a movement that spread across the world, a testament to the power of collective action in the face of a global crisis.

And so, the story of Climate Change became a tale of hope, a reminder that even in the darkest of times, there is always the possibility for change. The Earth may be in peril, but it's not too late to save it. The future is in our hands.

Chapter 2:

The Causes of Climate Change

Sarah, Jack, and Olivia continued their journey, seeking to understand the root causes of Climate Change. They visited scientists, academics, and experts in various fields, gathering information and insights on the issue.

One of the first things they learned was that Climate Change was caused by a buildup of greenhouse gases in the atmosphere. These gases, such as carbon dioxide, trapped heat from the sun and caused the Earth's temperature to rise.

"This is a problem that has been building for centuries," said a renowned climatologist they met. "Ever since the Industrial Revolution, humans have been burning fossil fuels like coal, oil, and natural gas, releasing massive amounts of carbon dioxide into the atmosphere."

"And it's not just the burning of fossil fuels," added a sustainability expert. "Deforestation, agriculture, and other human activities are also contributing to the problem. All of these things are making the Earth warmer, and it's having a devastating impact on our planet."

Sarah, Jack, and Olivia were shocked by what they heard. They had never realized the extent to which human activities were responsible for Climate Change.

"We have to do something," said Jack. "We have to stop this from getting worse."

"And we have to start now," added Olivia. "We can't afford to wait."

As they continued their journey, the trio encountered a group of skeptics who claimed that Climate Change was a natural phenomenon, not a human-made one. But after hearing from the experts, they were convinced that the overwhelming evidence pointed to human activities as the primary cause of the crisis.

"It's time for us to take responsibility for our actions and make the changes necessary to protect our planet," said Sarah.

The trio returned home with a deeper understanding of the causes of Climate Change, and a determination to spread the word and inspire others to take action. They knew that the future of the Earth depended on it.

As Sarah, Jack, and Olivia delved further into the topic of Climate Change, they began to grasp the enormity of the problem they were facing. They learned that the Earth's temperature was rising faster than at any time in the past 65 million years, and that the consequences of this were already being felt around the world.

They visited areas that were suffering from the impacts of Climate Change, such as rising sea levels, more frequent and intense hurricanes, and more severe droughts. They spoke to people who were losing their homes, their livelihoods, and even their lives as a result of the changing climate.

"It's hard to imagine that our actions are causing all of this," said Jack, as he surveyed the damage from a particularly severe hurricane. "But the facts are clear. We are the ones responsible for the Earth's temperature rising, and we have to take action to fix it."

Sarah, Jack, and Olivia soon discovered that there was hope. They learned about the numerous solutions that were available, from transitioning to renewable energy sources to reducing carbon emissions to planting trees. They met people who were already making a difference, working to create a more sustainable future for the Earth.

"We have the technology and the know-how to solve this problem," said Olivia. "All we need is the will to do it."

As the three friends returned home, they were filled with a sense of purpose. They knew that the future of the Earth was in their hands, and they were determined to use their voices to make a difference. They would use the lessons they had learned on their journey to inspire others to take action, and to do their part to ensure that the heat death of the Earth would never come to pass.

"We have the power to change the course of the future," said Sarah. "Let's use it wisely."

Chapter 3:

The Consequences of Climate Change

Sarah, Jack, and Olivia continued their mission to spread the word about Climate Change and its consequences. They were horrified by what they had seen and learned, and they were determined to make others see the urgency of the situation.

They started by visiting their local community, speaking to people about the effects of Climate Change and what could be done to prevent it. They organized events and talks, sharing their knowledge and insights with others. They also joined forces with local environmental groups, working together to raise awareness and advocate for change.

Their efforts paid off, and soon the community was buzzing with talk about the issue. People were becoming more and more concerned about the impacts of Climate Change, and they were eager to learn what they could do to help.

"We're finally getting through to people," said Jack, as he and Sarah watched a group of schoolchildren plant trees in a local park. "They're starting to understand just how important this issue is."

As they continued their mission, Sarah, Jack, and Olivia learned more about the consequences of Climate Change. They learned that it was causing more frequent and intense heatwaves, which were leading to widespread drought and famine. They learned that it was causing the polar ice caps to melt, which was causing sea levels to rise and putting coastal communities at risk. They also learned that it was causing more severe hurricanes, which were devastating communities and wiping out entire ecosystems.

"This is an emergency," said Olivia, as she and Sarah surveyed the aftermath of a particularly devastating hurricane. "We have to act fast, before it's too late."

The three friends were more determined than ever to make a difference. They knew that time was running out, and that every day mattered. They were resolved to use their voices to raise awareness, to inspire others to take action, and to do everything in their power to prevent the heat death of the Earth.

"We have a responsibility to future generations," said Sarah. "We can't let them down."

As their mission continued, Sarah, Jack, and Olivia encountered many people who were skeptical of Climate Change and its consequences. Some people simply didn't believe that the Earth was in such dire straits, while others believed that the issue was being exaggerated for political gain.

"It's frustrating," said Jack, as he and Olivia discussed the issue over coffee. "People just don't seem to get it. They don't realize the kind of world they're leaving behind for future generations."

Sarah, Jack, and Olivia decided that they needed to take their message directly to the people who were most skeptical. They knew that they had to engage with these people on a personal level, and to show them the facts and the evidence. They also knew that they had to be patient and understanding, and that they had to be willing to listen to their concerns and address their fears.

They started by reaching out to climate deniers and skeptics, inviting them to attend their events and talks. They engaged in debates and discussions, and they made sure to listen to their perspectives. They also shared with them the latest scientific studies and research, and they showed them the real-world impacts of Climate Change.

Over time, Sarah, Jack, and Olivia noticed a change in the attitudes of these people. They saw them starting to understand the urgency of the situation, and they saw them becoming more willing to take action.

"It's working," said Sarah, as she watched a group of skeptical business leaders signing a pledge to reduce their carbon emissions. "People are starting to realize just how critical this issue is."

As their mission continued, Sarah, Jack, and Olivia were more inspired than ever to fight for a better future. They were determined to spread the word, to mobilize people to action, and to do everything in their power to prevent the heat death of the Earth.

"We're not giving up," said Olivia, as she, Sarah, and Jack stood on a hill overlooking a pristine wilderness. "We're going to keep fighting, until this planet is safe for future generations."

With renewed vigor, Sarah, Jack, and Olivia continued to spread their message of hope and action. They spoke at conferences and events, they reached out to politicians and policymakers, and they worked with communities to raise awareness about the importance of reducing carbon emissions and protecting the environment.

One day, they received an invitation from the United Nations to present their findings and recommendations to a panel of experts and world leaders. It was a huge honor, and an opportunity for them to have a global impact.

At the UN meeting, Sarah, Jack, and Olivia presented their research and findings, and they made a passionate call for immediate action. They spoke about the need for renewable energy, for reducing carbon emissions, and for preserving natural habitats and wildlife. They spoke about the importance of investing in sustainable solutions, and of coming together as a global community to face this challenge.

Their message was received with applause and a standing ovation. The world leaders and experts were moved by their passion and commitment, and they agreed to work together to implement their recommendations.

From that day forward, Sarah, Jack, and Olivia became known as the Climate Change Warriors. They continued to inspire and motivate people around the world, and their work had a profound impact on the way that people thought about the environment and the future of our planet.

As the years passed, the Earth began to heal. The air became cleaner, the water clearer, and the wildlife flourished. The world became a better place, and future generations would never forget the courage and determination of Sarah, Jack, and Olivia, and their quest to prevent the heat death of the Earth.

Chapter 4:

"The Legacy Continues"

Sarah, Jack, and Olivia had accomplished more than they ever could have imagined. They had successfully brought the issue of climate change to the forefront of the world's attention and inspired millions of people to take action.

Their work, however, was far from over. The fight against climate change was an ongoing battle, and they knew that they had to continue to work hard to ensure that the progress they had made was not lost.

They continued to lead the charge, speaking at events and conferences, working with communities and policymakers, and advocating for renewable energy and sustainable solutions. They also founded the Climate Change Warriors Foundation, an organization dedicated to educating and empowering the next generation of environmental activists.

The Foundation quickly gained traction, and it soon had branches in cities and countries around the world. The Foundation provided training and resources to young activists, and it helped them to make their voices heard.

Years passed, and the Foundation grew stronger. The Climate Change Warriors became legends, and their story was told and retold to generations of young activists. The legacy of Sarah, Jack, and Olivia lived on, inspiring new generations to continue the fight against climate change and to work towards a better future for all.

As the world continued to heal, the Climate Change Warriors remained vigilant, never letting up in their efforts to protect the planet. They knew that their work was far from over, but they also knew that they had made a difference, and that the future was bright.

Sarah, Jack, and Olivia continued to lead the way in the fight against climate change, but they also made sure to pass on their knowledge and expertise to the next generation of activists. They knew that the fight was far from over and that there would always be new challenges and obstacles to overcome.

One day, they received a letter from a young girl named Emily, who had been inspired by their story and wanted to get involved in the fight against climate change. Sarah, Jack, and Olivia were touched by Emily's enthusiasm and passion, and they invited her to join the Climate Change Warriors Foundation.

Emily proved to be a quick learner, and she quickly became a valuable member of the team. She helped to organize events and rallies, and she spoke out against the use of fossil fuels and the destruction of natural habitats.

With Emily's help, the Foundation continued to grow and gain strength. It became a global movement, with branches in cities and countries all over the world. The Climate Change Warriors continued to inspire and motivate people to take action, and their message of hope and action resounded across the planet.

As the years passed, Sarah, Jack, and Olivia retired from the front lines of the fight against climate change. They took comfort in knowing that the work they had started would continue, and that their legacy would live on through the Foundation and through the countless activists who had been inspired by their story.

Chapter 5:

"The Future is Now"

Years passed, and the world continued to change. The use of renewable energy became widespread, and the emission of greenhouse gases declined. The world's leaders finally recognized the threat of climate change, and they took action to reduce their carbon footprints and protect the planet.

Emily, who had once been inspired by Sarah, Jack, and Olivia, was now a leader in her own right. She continued to work tirelessly to promote sustainability and to educate others about the importance of taking action on climate change.

One day, as she was walking through a park, she saw a group of children playing and laughing. They were playing a game that involved saving the planet from the effects of climate change, and they were having a great time. Emily smiled, knowing that the future was in good hands.

As she walked home, she thought about all the progress that had been made and all the work that still needed to be done. She knew that there was still much to be done, but she also knew that the world was on the right track.

The future was bright, and it was up to the next generation to continue the work of Sarah, Jack, and Olivia. The fight against climate change was far from over, but with the help of committed activists like Emily, it was a fight that could be won.

As the world becomes increasingly inhospitable, a growing number of people begin to question the wisdom of doing nothing. They start to organize, to protest, to demand action. At first, these efforts are small and fragmented, but as the years go by and the situation worsens, they become more unified and more influential.

In the midst of this growing movement, a charismatic young woman named Maya steps forward. She is passionate, articulate, and relentless in her efforts to raise awareness about the dangers of climate change. Through her tireless activism and her powerful speeches, she inspires countless others to join the cause and to fight for a better future.

Together, this growing movement of concerned citizens begins to make a real impact. They successfully pressure governments and corporations to take action, and they create a new sense of urgency and momentum in the fight against climate change.

And yet, even as they make progress, the challenges they face remain immense. The heat continues to rise, the storms become more frequent and more intense, and the impacts of climate change become increasingly devastating. But through it all, Maya and her fellow activists refuse to give up. They remain determined to save the planet, no matter the cost.

As the story comes to a close, the reader is left with a sense of hope and a sense of the power of collective action. Despite the odds, despite the challenges, the fight against climate change is far from over. And with the help of people like Maya, it may yet be won.

As the world becomes increasingly inhospitable, a growing number of people begin to question the wisdom of doing nothing. They start to organize, to protest, to demand action. At first, these efforts are small and fragmented, but as the years go by and the situation worsens, they become more unified and more influential.

In the midst of this growing movement, a charismatic young woman named Maya steps forward. She is passionate, articulate, and relentless in her efforts to raise awareness about the dangers of climate change. Through her tireless activism and her powerful speeches, she inspires countless others to join the cause and to fight for a better future.

Together, this growing movement of concerned citizens begins to make a real impact. They successfully pressure governments and corporations to take action, and they create a new sense of urgency and momentum in the fight against climate change.

And yet, even as they make progress, the challenges they face remain immense. The heat continues to rise, the storms become more frequent and more intense, and the impacts of climate change become increasingly devastating. But through it all, Maya and her fellow activists refuse to give up. They remain determined to save the planet, no matter the cost.

As the story comes to a close, the reader is left with a sense of hope and a sense of the power of collective action. Despite the odds, despite the challenges, the fight against climate change is far from over. And with the help of people like Maya, it may yet be won.

As the movement gains momentum, Maya and her team of activists begin to make real progress. They lead marches and rallies, they engage in civil disobedience, they stage protests outside the offices of the largest polluters. Slowly but surely, they begin to force the issue of climate change into the national discourse.

As the world becomes more aware of the urgent threat posed by climate change, many corporations and governments begin to take notice. They start to invest in renewable energy, to reduce their carbon emissions, and to support initiatives aimed at slowing the pace of global warming.

Despite these positive developments, however, there is still much work to be done. The global community is far from united in its efforts to combat climate change, and many powerful interests continue to resist meaningful action.

Maya and her team must navigate this complex and often hostile landscape, using all of their skills and all of their determination to build a better future. They must work to inspire and mobilize people everywhere, to create a global movement that can effect real change.

In the end, the outcome is uncertain. But Maya and her team remain unwavering in their resolve. They know that the stakes are too high, and that the future of the planet depends on the actions they take today.

With each passing day, the situation grows more dire, but Maya and her team remain hopeful. They know that with enough hard work and enough determination, they can create a world that is safe from the ravages of climate change. And so, with hearts full of hope and with spirits unbroken, they forge ahead, determined to make a difference, determined to save the planet.

Chapter 6

"The Power of Persistence: Maya and her Team's Journey to Inspire Climate Action"

As the months pass, Maya and her team of activists continue their tireless work. They travel the world, speaking to people in every corner of the globe, spreading their message and inspiring others to take action. They hold massive rallies, they stage sit-ins, they lead marches and protests.

Despite their efforts, however, there are still many who resist their message. Some dismiss climate change as a hoax, while others cling to outdated ideas and resist progress. Some see the issue as too complex or too distant to be a real concern, while others simply don't believe that their individual actions can make a difference.

Undeterred, Maya and her team continue their work, using every tool at their disposal to educate and inspire the public. They use social media and the internet to spread their message, they collaborate with scientists and experts to build a body of evidence that supports their cause.

And as the years go by, their message begins to take hold. Slowly but surely, people all around the world begin to realize the urgency of the crisis and the need for immediate action. They begin to demand that their governments take bold steps to address the problem, and they begin to make changes in their own lives that help reduce their carbon footprint.

Meanwhile, the world continues to warm, and the impacts of climate change become increasingly apparent. Coastal cities are flooded, deserts expand, and extreme weather events become more frequent.

Despite these challenges, however, Maya and her team remain steadfast in their commitment to creating a better future. They know that the battle will be long and difficult, but they also know that the outcome is far from predetermined. With enough hard work and enough determination, they believe that they can create a world that is safe from the ravages of climate change.

And so, with each passing day, they continue their work, determined to make a difference, determined to create a better future for all of us.

Maya and her team don't give up. They keep pushing for change, and as the years go by, their message begins to make a real impact. More and more people start to understand the urgency of the crisis and start to demand action from their governments. Companies and corporations, which once resisted change, start to realize that going green is not only good for the planet but also good for business. They start to invest in renewable energy, develop new technologies, and implement more sustainable practices.

People around the world start to see that change is possible, that they have the power to make a difference, and that everyone has a role to play in addressing the crisis. They start to see that it's not just about personal sacrifice but also about personal empowerment. They start to see that they can make a positive impact just by making simple changes in their daily lives, such as reducing their energy consumption, driving less, eating a more plant-based diet, and choosing products and services that are more environmentally friendly.

Despite these positive developments, the crisis continues to deepen, and the world continues to warm. But Maya and her team remain hopeful. They know that the road ahead will be long and challenging, but they also know that change is possible. They know that it's never too late to act and that every small step counts.

And so, they keep pushing, they keep inspiring, they keep leading the way. They know that their work may take many years, even decades, to bear fruit, but they also know that the future of the planet is at stake. And so, they keep fighting, driven by a fierce determination to create a better future for all of us.

In the end, it all comes down to one simple idea: that every single one of us has the power to make a difference. Whether we act or not, whether we choose to ignore the crisis or to face it head-on, will determine the fate of the planet and the future of humanity. The choice is ours, and the time is now.

As time passes, Maya and her team of activists continue their tireless efforts to spread their message about climate change. They travel the world, organizing rallies, protests, and sit-ins, and use social media and the internet to educate and inspire people to take action. Despite their hard work, there are still many who resist their message, and some still deny the reality of climate change. However, Maya and her team continue to collaborate with scientists and experts, using evidence-based arguments to support their cause. Over the years, their message starts to gain traction, and more people begin to realize the urgent need for action on climate change. They start to demand that their governments take bold steps to address the issue, and individuals begin to make changes in their daily lives to reduce their carbon footprint. Even companies and corporations start to realize that going green is not only good for the planet but also good for business, and they start to invest in renewable energy and sustainable practices. Despite the progress made, the effects of climate change are still becoming more evident. Coastal cities are flooded, deserts expand, and extreme weather events become more frequent. However, Maya and her team remain determined to create a better future for all. They continue to push for change and inspire others to join them in the fight against climate change. They remain hopeful and steadfast in their commitment to the cause, knowing that every small step counts in the fight to create a safer future. In the end, the battle against climate change comes down to one fundamental idea: that every individual has the power to make a difference. The choice to act or ignore the issue ultimately determines the fate of the planet and humanity's future. Maya and her team continue to work tirelessly, driven by a fierce determination to create a better future for all of us. Even though the road ahead is long and challenging, they remain hopeful that change is possible, and that their work will bear fruit in the years and decades to come.

Conclusion

Maya and her team's tireless work has not gone in vain. Despite the resistance, they continued to inspire people around the world to take action and demand change. Their message has gradually taken hold, and more and more individuals, corporations, and governments have started to prioritize climate action and adopt sustainable practices.

However, the crisis of climate change continues to deepen, and the planet's future remains at stake. It's up to every one of us to take action and make a positive impact, whether through personal choices or advocacy and activism. The road ahead may be long and challenging, but with determination, cooperation, and innovative solutions, a better future for all of us is possible.

In the end, the book's message is clear: the power to make a difference lies within each one of us. The time for action is now, and the fate of the planet and humanity is in our hands.

www.ingramcontent.com/pod-product-compliance
Lightning Source LLC
Chambersburg PA
CBHW071147220526
45467CB00015B/2101